The Open University

Mathematics: A Second Level Course

Linear Mathematics Unit 16

EUCLIDEAN SPACES I: INNER PRODUCTS

Prepared by the Linear Mathematics Course Team

The Open University Press

The Open University Press Walton Hall Bletchley Bucks

First published 1972

Designed by the Media Development Group of the Open University

Printed in Great Britain by
Martin Cadbury Printing Group

SBN 335 01108 X

This text forms part of the correspondence element of an Open University Second Level Course. The complete list of units in the course is given at the end of this text.

For general availability of supporting material referred to in this text, please write to the Director of Marketing, The Open University, Walton Hall, Bletchley, Buckinghamshire.

Further information on Open University courses may be obtained from The Admissions Office, The Open University, P.O. Box 48, Bletchley, Buckinghamshire.

1.1

Contents

Set Books

D. L. Kreider, R. G. Kuller, D. R. Ostberg and F. W. Perkins, *An Introduction to Linear Analysis* (Addison-Wesley, 1966).

E. D. Nering, *Linear Algebra and Matrix Theory* (John Wiley, 1970).

It is essential to have these books; the course is based on them and will not make sense without them.

Conventions

Before working through this correspondence text make sure you have read *A Guide to the Linear Mathematics Course.* Of the typographical conventions given in the Guide the following are the most important.

The set books are referred to as:

K for *An Introduction to Linear Analysis*
N for *Linear Algebra and Matrix Theory*

All starred items in the summaries are examinable.

References to the Open University Mathematics Foundation Course Units (The Open University Press, 1971) take the form *Unit M100 3, Operations and Morphisms.*

16.0 INTRODUCTION

In this unit we look at a new type of space with a structure richer than vector space structure. The new space is called a *Euclidean space*, and the way it differs from a vector space can be stated in geometrical terms.

In a vector space we do not define the length of a vector, nor do we define the angle between two vectors. (If one vector is a scalar multiple of another, then we can say that their lengths are in the ratio of this scalar, and that the angle between them is 0° (or 180°) but this is a very special case: if the vectors are linearly independent such comparisons make no sense.) Loosely, we could say that the language of vector spaces does not contain any words for "length" or "angle". What we want to do now is to add these words to our mathematical vocabulary; that is, we shall define a new mathematical structure which contains concepts that cannot be described in the "vector space language". You will remember that we describe a vector space as a set of objects (vectors) together with two operations: addition of vectors and multiplication of a vector by a scalar. To convert the vector space to a Euclidean space we attach one more operation to the list. The new operation is a new type of multiplication: the *inner product*. It is a binary operation on vectors that is not closed: the inner product of two vectors is a *scalar*, not a vector.

You have already met in *Unit M100 22, Linear Algebra I*, the inner product for geometric vectors:

$$\underset{\to}{x} \cdot \underset{\to}{y} = |\underset{\to}{x}|\ |\underset{\to}{y}| \cos\theta$$

where $\underset{\to}{x}$ and $\underset{\to}{y}$ are geometric vectors, $|\underset{\to}{x}|$ and $|\underset{\to}{y}|$ their lengths, and θ the angle between them.

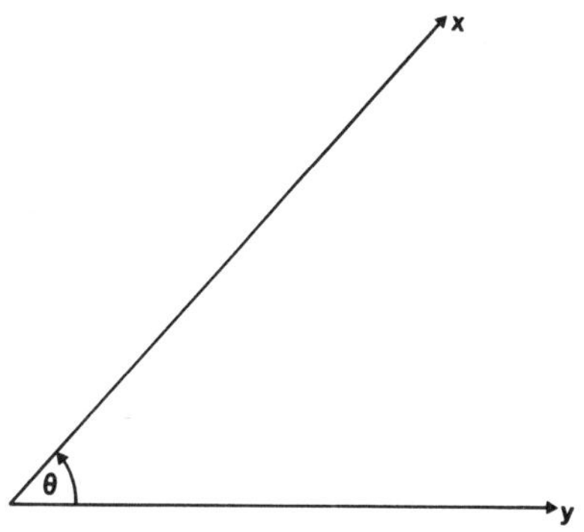

Notice that $\underset{\to}{x} \cdot \underset{\to}{y}$ is a scalar, not a vector. If we use this operation as well as the two vector space operations, we are using geometric vectors as a Euclidean space. For example, we can use it to tell us when two vectors are at right-angles; for if they are we have $\theta = \pi/2$ and so $\cos\theta = 0$, and hence $\underset{\to}{x} \cdot \underset{\to}{y} = 0$. Thus the inner product allows us to make statements about angles in algebraic terms which we could not do using vector addition and scalar multiplication alone.

16.1 INNER PRODUCTS

16.1.1 Definition

To define an inner product by generalizing the formula $\underset{\sim}{x} \cdot \underset{\sim}{y} = |\underset{\sim}{x}|\ |\underset{\sim}{y}| \cos \theta$ would be to beg the question—we do not yet know how to define the angle θ or the length $|\mathbf{x}|$. The method we adopt is in fact the exact opposite. We define $\mathbf{x} \cdot \mathbf{y}$ first, then define length and angle in terms of it. The definition of $\mathbf{x} \cdot \mathbf{y}$ is formulated in the first reading passage.

READ from page **K**256 *to the line above Example* 1 *on page* **K**257.

Note

Last paragraph of reading passage The Equations (7-2), (7-3) and (7-5) tell us that the inner product is a bilinear form as defined in *Unit 14, Bilinear and Quadratic Forms.* This is also exactly the condition of Equation (7-6). From Equation (7-1) we know that this bilinear form has to be symmetric and Equation (7-4) tells us that the associated quadratic form is positive definite. Hence we could have defined an inner product function as a real symmetric positive definite bilinear form.

To determine whether a suggested inner product is indeed an inner product, we must check that

$\mathbf{x} \cdot \mathbf{y}$ is defined on each pair of vectors;

$\mathbf{x} \cdot \mathbf{y}$ is real;

each of properties (7-1), (7-2), (7-3) and (7-4) holds.

Example

The inner product for geometric vectors in R^3 satisfies each of the above points*. (You can verify this by referring to sub-section 22.1.6 of *Unit M100 22, Linear Algebra I.*)

Exercises

1. Exercise 11, page **K**260.
2. Exercise 7, page **K**259. (Read the "whenever" in Equation (7-7) as "if".)
3. If $\mathbf{x}_1$, $\mathbf{x}_2$, $\mathbf{x}_3$ are such that $\mathbf{x}_i \cdot \mathbf{x}_j = 0$ for $i \neq j$, and

$$\mathbf{x} = \alpha_1 \mathbf{x}_1 + \alpha_2 \mathbf{x}_2 + \alpha_3 \mathbf{x}_3 = \sum_{i=1}^{3} \alpha_i \mathbf{x}_i,$$

$$\mathbf{y} = \beta_1 \mathbf{x}_1 + \beta_2 \mathbf{x}_2 + \beta_3 \mathbf{x}_3 = \sum_{i=1}^{3} \beta_i \mathbf{x}_i,$$

find γ_i such that

$$\mathbf{x} \cdot \mathbf{y} = \sum_{i=1}^{3} \gamma_i \mathbf{x}_i \cdot \mathbf{x}_i .$$

Solutions

1. (a) Since $\mathbf{x} \cdot \mathbf{y}$ is an inner product, $\mathbf{x} \circ \mathbf{y}$ is real and defined for all $\mathbf{x}$, $\mathbf{y}$.

We verify conditions (7-1) to (7-4).

(7-1) $\quad \mathbf{x} \circ \mathbf{y} = 2(\mathbf{x} \cdot \mathbf{y}) = 2(\mathbf{y} \cdot \mathbf{x}) = \mathbf{y} \circ \mathbf{x}$

(7-2) $\quad (\alpha\mathbf{x}) \circ \mathbf{y} = 2(\alpha\mathbf{x}) \cdot \mathbf{y}) = 2\alpha(\mathbf{x} \cdot \mathbf{y}) = \alpha(\mathbf{x} \circ \mathbf{y})$

* We shall use R^2 and R^3 to label the spaces of geometric vectors. See Section 1–8 of **K** for a discussion of the isomorphism of the two-dimensional case.

(7-3) $$(\mathbf{x}_1 + \mathbf{x}_2) \circ \mathbf{y} = 2((\mathbf{x}_1 + \mathbf{x}_2) \cdot \mathbf{y}) = 2(\mathbf{x}_1 \cdot \mathbf{y} + \mathbf{x}_2 \cdot \mathbf{y}) = 2(\mathbf{x}_1 \cdot \mathbf{y}) + 2(\mathbf{x}_2 \cdot \mathbf{y}) = \mathbf{x}_1 \circ \mathbf{y} + \mathbf{x}_2 \circ \mathbf{y}$$

(7-4) $\mathbf{x} \circ \mathbf{x} = 2(\mathbf{x} \cdot \mathbf{x}) \geqslant 0$ as $\mathbf{x} \cdot \mathbf{x} \geqslant 0$.
$0 = \mathbf{x} \circ \mathbf{x} = 2(\mathbf{x} \cdot \mathbf{x})$ if and only if $\mathbf{x} \cdot \mathbf{x} = 0$,

i.e. if and only if $\mathbf{x} = \mathbf{0}$.

(b) For any α we are able to verify conditions (7-1), (7-2), (7-3). However, for condition (7-4) to hold it is necessary and sufficient that $\alpha > 0$.

2. We start by proving $\mathbf{0} \cdot \mathbf{y} = 0$ for all $\mathbf{y}$. Following the hint we consider $(\mathbf{0} + \mathbf{0}) \cdot \mathbf{y}$. Expanding, we obtain

$$(\mathbf{0} + \mathbf{0}) \cdot \mathbf{y} = \mathbf{0} \cdot \mathbf{y} + \mathbf{0} \cdot \mathbf{y}$$

Also $(\mathbf{0} + \mathbf{0}) \cdot \mathbf{y} = \mathbf{0} \cdot \mathbf{y}$, as $\mathbf{0} + \mathbf{0} = \mathbf{0}$.

Hence $\mathbf{0} \cdot \mathbf{y} = \mathbf{0} \cdot \mathbf{y} + \mathbf{0} \cdot \mathbf{y}$.

Subtracting $\mathbf{0} \cdot \mathbf{y}$ from both sides we have $0 = \mathbf{0} \cdot \mathbf{y}$ as required.

Since the inner product is symmetric we prove $\mathbf{x} \cdot \mathbf{0} = \mathbf{0} \cdot \mathbf{x} = 0$ by the above with $\mathbf{x}$ for $\mathbf{y}$. Note, however, that the converse is not true: if $\mathbf{x} \cdot \mathbf{y} = 0$ it does not follow that either $\mathbf{x}$ or $\mathbf{y}$ is zero. For example, with geometric vectors, we can have $\cos \theta = 0$, i.e. $\mathbf{x}$ and $\mathbf{y}$ are at right angles.

3. By bilinearity we have that

$$\mathbf{x} \cdot \mathbf{y} = \left(\sum_{i=1}^{3} \alpha_i \mathbf{x}_i\right) \cdot \left(\sum_{j=1}^{3} \beta_j \mathbf{x}_j\right) = \sum_{i=1}^{3} \sum_{j=1}^{3} \alpha_i \beta_j \mathbf{x}_i \cdot \mathbf{x}_j .$$

The only non-zero terms occur when $i = j$. Hence

$$\mathbf{x} \cdot \mathbf{y} = \sum_{i=1}^{3} \alpha_i \beta_i \mathbf{x}_i \cdot \mathbf{x}_i = \alpha_1 \beta_1 \mathbf{x}_1 \cdot \mathbf{x}_1 + \alpha_2 \beta_2 \mathbf{x}_2 \cdot \mathbf{x}_2 + \alpha_3 \beta_3 \mathbf{x}_3 \cdot \mathbf{x}_3 .$$

i.e. $\gamma_i = \alpha_i \beta_i$, $i = 1, 2, 3$.

16.1.2 Examples

We now look at some very common inner products.

READ from Example 1 *on page* **K**257 *to the end of Section* 7-1.

Note

line 11, page **K**258 Note these conventions.

Examples

1. Exercise 1(b), page **K**259.

$$\mathbf{x} = (\tfrac{2}{3}, \tfrac{1}{2}, 1); \ \mathbf{y} = (-\tfrac{1}{2}, 4, 2).$$

We use the definition of $\mathbf{x} \cdot \mathbf{y}$ in $\mathscr{R}^n$, given in Equation (7-8) on page **K**257,

$$\mathbf{x} \cdot \mathbf{y} = \tfrac{2}{3} \times (-\tfrac{1}{2}) + \tfrac{1}{2} \times 4 + 1 \times 2 = -\tfrac{1}{3} + 2 + 2 = 3\tfrac{2}{3}.$$

2. Exercise 2(b), page **K**259.

From Equation (7-9)

$$\mathbf{f}\cdot\mathbf{g} = \int_0^1 x(1-x)\,dx$$

$$= \int_0^1 (x - x^2)\,dx$$

$$= \tfrac{1}{2} - \tfrac{1}{3} = \tfrac{1}{6}.$$

3. Find $\mathbf{f}\cdot\mathbf{g}$ for the following vectors in $C[0, 1]$ when the inner product is defined with respect to the weight function $\mathbf{r}(x) = e^x$.

$$\mathbf{f}(x) = xe^{-x/2};\ \mathbf{g}(x) = (1+x)e^{-x/2}.$$

By Equation (7-10), we have

$$\mathbf{f}\cdot\mathbf{g} = \int_0^1 xe^{-x/2}(1+x)e^{-x/2}e^x\,dx$$

$$= \int_0^1 (x + x^2)\,dx = \tfrac{5}{6}.$$

Exercises

1. Exercise 1(a), page **K**259.
2. Exercise 2(a), page **K**259.
3. Exercise 3(a), page **K**259.
4. Exercise 4, page **K**259.
5. Given any 2×2 matrix A we can define a bilinear form on the space of 2×1 matrices as follows

$$(X, Y) \longmapsto X\cdot Y = X^T A Y.$$

For which of the following matrices is this an inner product?

(a) $\begin{bmatrix} 2 & 1 \\ 1 & 1 \end{bmatrix}$, (b) $\begin{bmatrix} -1 & 0 \\ 1 & 0 \end{bmatrix}$, (c) $\begin{bmatrix} 1 & 1 \\ 1 & 1 \end{bmatrix}$.

(*Hint:* an inner product is a real symmetric positive definite bilinear form.)

6. Exercise 10, page **K**260.

Solutions

1. $\mathbf{x}\cdot\mathbf{y} = (\tfrac{1}{2}, 2, -1)\cdot(4, -2, 3)$
$= \tfrac{1}{2}\times 4 - 2\times 2 - 1\times 3 = 2 - 4 - 3 = -5.$

2. $$\mathbf{f}\cdot\mathbf{g} = \int_0^1 x(1-x^2)\,dx$$
$$= \int_0^1 x\,dx - \int_0^1 x^3\,dx$$
$$= \tfrac{1}{2} - \tfrac{1}{4} = \tfrac{1}{4}.$$

3. $$\mathbf{f}\cdot\mathbf{g} = \int_0^1 e^x e^{-x}(1-2x)\,dx = [x - x^2]_0^1 = 0.$$

4. $\mathbf{x}\cdot\mathbf{y}$ is real and defined for all $\mathbf{x}$, $\mathbf{y}$.

We verify the conditions (7-1) to (7-4).

(7-1) $\mathbf{x}\cdot\mathbf{y} = x_1y_1 + \cdots + x_ny_n = y_1x_1 + \cdots + y_nx_n$
$= \mathbf{y}\cdot\mathbf{x}$

(7-2) $(\alpha\mathbf{x})\cdot\mathbf{y} = \alpha x_1y_1 + \alpha x_2y_2 + \cdots + \alpha x_ny_n$
$= \alpha(x_1y_1 + \cdots + x_ny_n) = \alpha(\mathbf{x}\cdot\mathbf{y})$

(7-3) $$\begin{aligned}(\mathbf{x}+\mathbf{x}')\cdot\mathbf{y} &= (x_1+x_1')y_1+\cdots+(x_n+x_n')y_n\\ &= x_1y_1+\cdots+x_ny_n+x_1'y_1\\ &\quad+\cdots+x_n'y_n\\ &= \mathbf{x}\cdot\mathbf{y}+\mathbf{x}'\cdot\mathbf{y}\end{aligned}$$

(7-4) Since $\mathbf{x}\cdot\mathbf{x}=\sum_{i=1}^{n} x_i^2, \mathbf{x}\cdot\mathbf{x}\geqslant 0$, and $\mathbf{x}\cdot\mathbf{x}=0$ if and only if $x_i=0$ for all i, that is $\mathbf{x}=\mathbf{0}$.

5. In each case we have a real bilinear form so we only need to check that it is symmetric and positive definite.

(a) Since $\begin{bmatrix}2 & 1\\ 1 & 1\end{bmatrix}$ is symmetric, the bilinear form is symmetric.

Also, if $X=\begin{bmatrix}x_1\\ x_2\end{bmatrix}$,

$$\begin{aligned}X\cdot X &= 2x_1^2+2x_1x_2+x_2^2\\ &= x_1^2+(x_1+x_2)^2\end{aligned}$$

i.e. $X\cdot X\geqslant 0$ and $X\cdot X=0$ if and only if $X=\begin{bmatrix}0\\ 0\end{bmatrix}$.

Thus the bilinear form is positive definite and hence defines an inner product.

(b) Since $\begin{bmatrix}-1 & 0\\ 1 & 0\end{bmatrix}$ is not symmetric, it does not define an inner product.

(c) If $X=\begin{bmatrix}1\\ -1\end{bmatrix}$ and $Y=\begin{bmatrix}1\\ -1\end{bmatrix}$

$$X^TAY=0,$$

and hence the matrix does not define an inner product.

6. (a) $\mathbf{p}\cdot\mathbf{q}$ is defined for each pair of vectors and $\mathbf{p}\cdot\mathbf{q}$ is real.

(7-1) follows since multiplication on the reals is commutative.

(7-2) follows since multiplication on the reals is distributive over addition.

(7-3) follows since multiplication on the reals is distributive over addition.

(7-4) Since $\mathbf{p}\cdot\mathbf{p}=\sum_{i=0}^{n} a_i^2$, $\mathbf{p}\cdot\mathbf{p}\geqslant 0$ and $\mathbf{p}\cdot\mathbf{p}=0$ if and only if $\mathbf{p}(x)=0+0x+\cdots+0x^n$, i.e. $\mathbf{p}=\mathbf{0}$.

(b) $\mathscr{P}$ as a vector space is a subspace of $\mathcal{C}[a,b]$ as a vector space. We therefore compare the definitions of the inner product given for $\mathscr{P}$ and $\mathcal{C}[a,b]$. The inner products do not look the same; we therefore investigate the definitions for two simple vectors.

If $\mathbf{p}(x)=x$ and $\mathbf{q}(x)=x^2$,

in $\mathscr{P}$, $\mathbf{p}\cdot\mathbf{q}=0$;

in $\mathcal{C}[a,b]$, $\mathbf{p}\cdot\mathbf{q}=\int_a^b x^3\,dx=\dfrac{b^4-a^4}{4}\neq 0$ if $b\neq -a$.

Hence $\mathscr{P}$ is not a subspace of $\mathcal{C}[a,b]$. Other examples show that $\mathscr{P}$ is not a subspace of $\mathcal{C}\,[a,b]$ for any a, b.

16.1.3 Summary of Section 16.1

In this section we defined the terms

inner product	(page **K**256)	* * *
Euclidean space	(page **K**256)	* * *
Euclidean *n*-space	(page **K**257)	* * *
weight function	(page **K**258)	* * *

The most important inner products for our purposes are:

$\mathbf{x} \cdot \mathbf{y} = \sum_{i=1}^{n} x_i y_i$ in R^n (page **K**257) * * *

$\mathbf{f} \cdot \mathbf{g} = \int_a^b f(x)g(x)r(x)\, dx$ in $C[a, b]$ (page **K**258) * * *

where $r(x)$ is non-negative with only a finite number of zeros in $[a, b]$.

Technique

Given two vectors in R^n or two functions in $C[a, b]$, find their inner product. * * *

16.2 LENGTH, ANGLE, DISTANCE

16.2.1 Definition of Length

READ Section 7-2 starting on page **K**261, *to line 4 on page* **K**262.

Examples

1. Exercise 1(b), page **K**265.

By Formula (7-13), the length of **x** is

$$\begin{aligned}\|\mathbf{x}\| &= \sqrt{(\tfrac{1}{2})^2 + (-\sqrt{3})^2 + (\sqrt{2})^2 + (1)^2}\\ &= \sqrt{6.25} = 2.5.\end{aligned}$$

2. Exercise 3(a), page **K**265.

By Formula (7-14), the length of **f** is

$$\begin{aligned}\|\mathbf{f}\| &= \left(\int_0^1 x \times x \, dx\right)^{1/2}\\ &= \left(\left[\frac{x^3}{3}\right]_0^1\right)^{1/2} = \left(\frac{1}{3}\right)^{1/2}.\end{aligned}$$

Exercises

1. Exercise 1(c), page **K**265.
2. Exercise 3(b), page **K**265.
3. Exercise 15, page **K**267.

Solutions

1. $$\begin{aligned}\|(3, 4, -3, 1)\| &= \sqrt{(3, 4, -3, 1) \cdot (3, 4, -3, 1)}\\ &= \sqrt{9 + 16 + 9 + 1}\\ &= \sqrt{35}.\end{aligned}$$

2. $$\begin{aligned}\|\mathbf{f}\| &= \left(\int_0^1 [f(x)]^2 \, dx\right)^{1/2} = \left(\int_0^1 (e^{x/2})^2 \, dx\right)^{1/2}\\ &= \left(\int_0^1 e^x \, dx\right)^{1/2} = (e - 1)^{1/2}.\end{aligned}$$

3. $\|\alpha\mathbf{x}\| = \sqrt{\alpha\mathbf{x} \cdot \alpha\mathbf{x}} = \sqrt{\alpha^2}\,\sqrt{\mathbf{x} \cdot \mathbf{x}} = |\alpha|\,\|\mathbf{x}\|.$

16.2.2 Angle and the Schwarz Inequality

How do we define *angle* in terms of inner products and lengths? In the Foundation Course we defined the inner product on geometric vectors by

$$\underset{\to}{x} \cdot \underset{\to}{y} = |\underset{\to}{x}|\,|\underset{\to}{y}| \cos\theta$$

and hence

$$\cos\theta = \frac{\underset{\to}{x} \cdot \underset{\to}{y}}{|\underset{\to}{x}|\,|\underset{\to}{y}|}$$

The expression $\dfrac{\mathbf{x} \cdot \mathbf{y}}{|\mathbf{x}|\,|\mathbf{y}|}$ is meaningful for any Euclidean space, and as we shall see in the next reading passage, enables us to define the angle between two vectors **x** and **y**.

READ from line 5 on page **K**262 *to the end of page* **K**263.

Notes

(i) *line 7, page* **K**262 The law of cosines states that if a, b and c are the lengths of the sides of a triangle and θ is the angle opposite the side of length a, then

$$a^2 = b^2 + c^2 - 2bc \cos \theta$$

(ii) *lines −8 and −1, page* **K**262 The following figures illustrate these two lines. In the left -hand figure α and β are chosen to make OP = OQ.

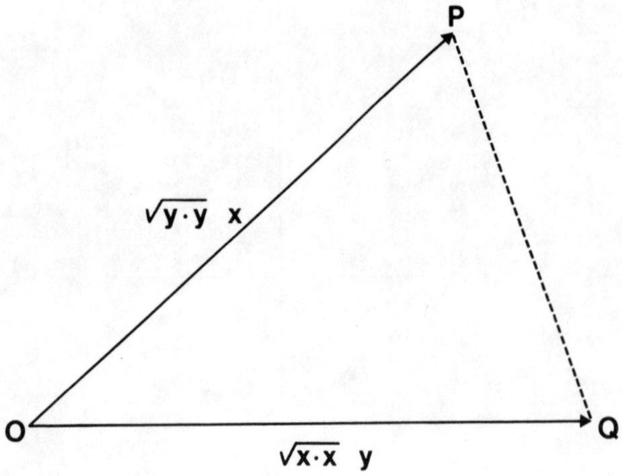

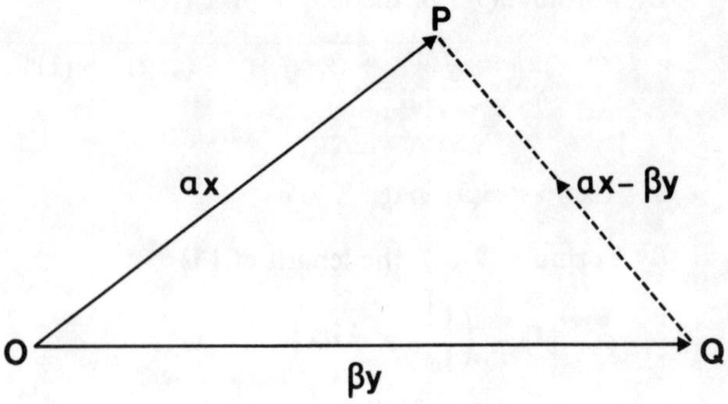

(iii) *line −2, page* **K**263 Therefore, the angle may be obtuse but not negative.

Example

Exercise 4(b), page **K**266.

$$\|\mathbf{x}\| = \sqrt{\tfrac{1}{4} + 1 + 4},\ \|\mathbf{y}\| = \sqrt{4 + 16 + 64},$$
$$\mathbf{x} \cdot \mathbf{y} = 1 + 4 + 16 = 21.$$

$$\text{Hence } \cos\theta = \frac{\mathbf{x}\cdot\mathbf{y}}{\|\mathbf{x}\|\ \|\mathbf{y}\|} = \frac{21}{\sqrt{\frac{21}{4}}\sqrt{84}} = 1$$

i.e. $\theta = 0$.

(Did you notice that $\mathbf{y} = 4\mathbf{x}$ and hence the vectors are linearly dependent?)

Exercises

1. Exercise 4(a), page **K**266.
2. Exercise 4(e), page **K**266.
3. Exercise 9, page **K**266.

Solutions

1. The angle between $\mathbf{x} = (1, 1, 1)$ and $\mathbf{y} = (\frac{1}{2}, -1, \frac{1}{2})$ is θ where

$$\cos\theta = \frac{\mathbf{x}\cdot\mathbf{y}}{\|\mathbf{x}\|\ \|\mathbf{y}\|} = 0.$$

Hence $\theta = \pi/2$: the two vectors are at right angles.

2. The angle between $\mathbf{x} = (-3, -1, 0)$ and $\mathbf{y} = (1, 2, -\sqrt{5})$ is θ where

$$\cos\theta = \frac{\mathbf{x}\cdot\mathbf{y}}{\|\mathbf{x}\|\ \|\mathbf{y}\|} = \frac{-5}{\sqrt{10}\sqrt{10}} = -\tfrac{1}{2}.$$

$$\text{Hence } \theta = \frac{2\pi}{3}.$$

3. Comparing the desired result

$$(a_1 + \cdots + a_n)\left(\frac{1}{a_1} + \cdots + \frac{1}{a_n}\right) \geqslant n^2$$

with Cauchy's inequality, Equation (7-18), suggests we look for vectors **x** and **y** in $\mathscr{R}^n$ such that

$$\sum_{i=1}^{n} x_i^2 = \sum_{i=1}^{n} a_i, \quad \text{and} \quad \sum_{i=1}^{n} y_i^2 = \sum_{i=1}^{n} \frac{1}{a_i}$$

So we try $\mathbf{x} = (\sqrt{a_1}, \ldots, \sqrt{a_n})$ and $\mathbf{y} = \left(\frac{1}{\sqrt{a_1}}, \ldots, \frac{1}{\sqrt{a_n}}\right)$,

and since these give

$$\mathbf{x} \cdot \mathbf{y} = \sum_{i=1}^{n} x_i y_i = 1 + \cdots + 1 = n,$$

the Cauchy inequality gives the result we want.

16.2.3 Distance

READ from page **K**264 *to the end of Section* 7-2 *on page* **K**265.

Notes

(i) *line −1, page* **K**264 The following figures illustrate Equation (7-27).

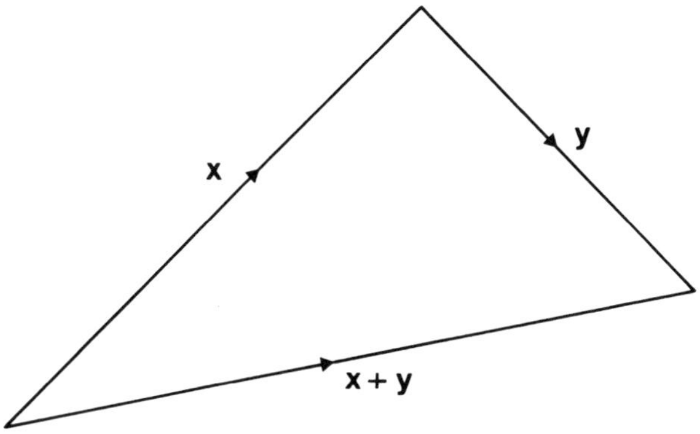

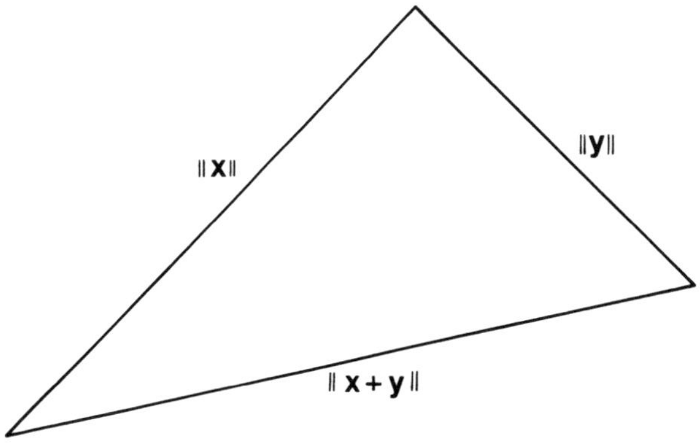

(ii) *line 10, page* **K**265 Equation (7-28) is illustrated by the following figure.

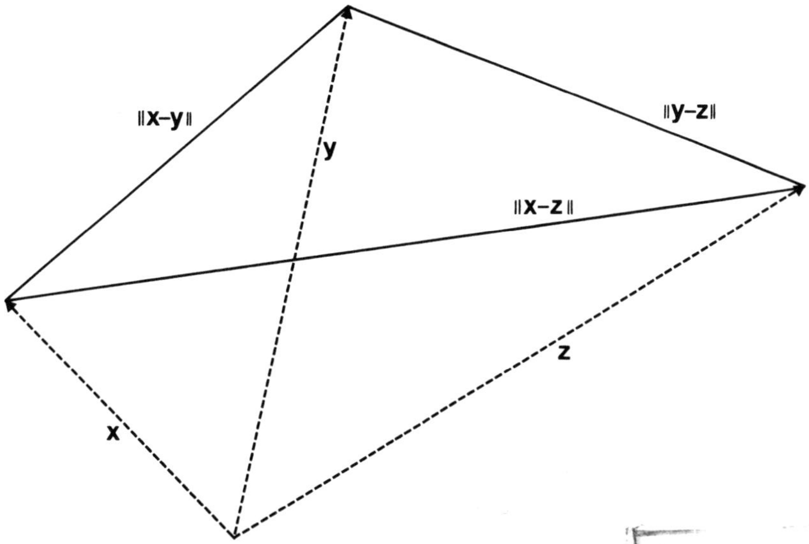

Example

Exercise 2(b), page **K**265.

We first calculate $\mathbf{x} - \mathbf{y}$:

$$(7, -4, 1, 3) - (2, 1, -4, 8) = (5, -5, 5, -5).$$

The distance between **x** and **y** is therefore:

$$\|\mathbf{x} - \mathbf{y}\| = \sqrt{25 + 25 + 25 + 25} = 10.$$

Exercise

Exercise 2(a), page **K**265.

Solution

$$\mathbf{x} - \mathbf{y} = (1, -2, 0, 2)$$

The distance between **x** and **y** is

$$\|\mathbf{x} - \mathbf{y}\| = \sqrt{1 + 4 + 0 + 4} = 3.$$

16.2.4 Discussion

We have seen how an *inner product* on a vector space can be used to define a *norm* and then *distance*. All these ideas have to some extent separate existences. There are spaces which have a norm but no inner product and spaces which have a distance but no norm. Whereas an *inner product* on a real vector space V is a mapping

$$V \times V \longrightarrow R$$

with the properties of Definition (7.1), a *norm* is a mapping

$$n: V \longrightarrow R, \text{ where } n(\mathbf{x}) = \|\mathbf{x}\|,$$

such that

(i) $\|\mathbf{x}\| = 0 \Leftrightarrow \mathbf{x} = \mathbf{0}$.

(ii) $\|a\mathbf{x}\| = |a| \; \|\mathbf{x}\| \qquad (a \in R)$

(iii) $\|\mathbf{x} + \mathbf{y}\| \leqslant \|\mathbf{x}\| + \|\mathbf{y}\|$

and a *distance* is a mapping

$$d: V \times V \longrightarrow R$$

such that

(i) $d(\mathbf{x}, \mathbf{y}) = 0 \Leftrightarrow \mathbf{x} = \mathbf{y}$

(ii) $d(\mathbf{x}, \mathbf{y}) \geqslant 0$

(iii) $d(\mathbf{x}, \mathbf{y}) = d(\mathbf{y}, \mathbf{x})$

(iv) $d(\mathbf{x}, \mathbf{y}) + d(\mathbf{y}, \mathbf{z}) \geqslant d(\mathbf{x}, \mathbf{z})$.

The distance axioms do not use any vector space properties and it is possible to define a distance function on an arbitrary set.

The relationship between these ideas is that given an inner product, we can define a norm $\|\mathbf{x}\| = \sqrt{\mathbf{x} \cdot \mathbf{x}}$, and given a norm, we can define a distance $d(x, y) = \|\mathbf{x} - \mathbf{y}\|$. These definitions do in fact give a well defined norm and distance. Most of the properties needed have been verified in the text.

16.2.5 Summary of Section 16.2

In this section we defined the terms

length (norm)	(page **K**261)	* * *
angle	(page **K**263)	* *
distance	(page **K**264)	* * *

Theorems

1. (**7-1,** page **K**262) (Schwarz inequality)
If $\mathbf{x}$ and $\mathbf{y}$ are any two vectors in a Euclidean space, then * * *

$$(\mathbf{x} \cdot \mathbf{y})^2 \leqslant (\mathbf{x} \cdot \mathbf{x})\,(\mathbf{y} \cdot \mathbf{y}).$$

2. (**Lemma 7-1,** page **K**264) (triangle inequality)
If $\mathbf{x}$ and $\mathbf{y}$ are arbitrary vectors in a Euclidean space, then * * *

$$\|\mathbf{x} + \mathbf{y}\| \leqslant \|\mathbf{x}\| + \|\mathbf{y}\|.$$

Technique

Given two vectors in a Euclidean space, find their lengths and the angle and distance between them. * * *

Notation

$\|\mathbf{x}\|$	(page **K**261)
$d(\mathbf{x}, \mathbf{y})$	(page **K**264)

16.3 ORTHOGONALITY

16.3.1 Definition

Having defined the angle between two vectors, we can look at the possibility of two vectors being perpendicular. In the case of geometric vectors, we say that two vectors are perpendicular if the angle between them is $\frac{\pi}{2}$, i.e. $\cos\theta = 0$. In the same way, two vectors in Euclidean space are perpendicular or *orthogonal* if $\cos\theta = 0$, i.e. if and only if $\mathbf{x} \cdot \mathbf{y} = 0$.

Example

If $\mathbf{x} = (1, 2, -1, -3)$ and $\mathbf{y} = (3, -2, -4, 1)$, $\mathbf{x} \cdot \mathbf{y} = 0$; i.e., $\mathbf{x}$ and $\mathbf{y}$ are orthogonal

READ from page **K**268 *to line −10, page* **K**270

Notes

(i) *Definition 7-5, page* **K**268 Note that the zero vector, **0**, is orthogonal to every vector but **0** cannot be in an orthogonal set.
(ii) *line 8, page* **K**269 A unit vector is a vector of length 1.
(iii) *Equation* (*7-37*), *page* **K**270 These integrals can be evaluated with the aid of Section III. 5.1 of **TI**.

Example

Find a vector of unit length which is orthogonal to $\mathbf{a} = (2, 1, 3)$ and $\mathbf{b} = (1, 1, 1)$.

Let $\mathbf{c} = (c_1, c_2, c_3)$ be a vector that is orthogonal to **a** and to **b**. Then

$$\mathbf{a} \cdot \mathbf{c} = 2c_1 + c_2 + 3c_3 = 0$$

and

$$\mathbf{b} \cdot \mathbf{c} = c_1 + c_2 + c_3 = 0$$

Reducing to Hermite normal form gives

$$c_1 + 2c_3 = 0$$
$$c_2 - c_3 = 0$$

so that $\mathbf{c} = (-2c_3, c_3, c_3)$.

The condition that **c** be of unit length gives

$$\|c\| = \sqrt{4c_3^2 + c_3^2 + c_3^2} = 1,$$

i.e. $c_3 = \pm\sqrt{\frac{1}{6}}$.

Thus both $\left(\frac{-2}{\sqrt{6}}, \frac{1}{\sqrt{6}}, \frac{1}{\sqrt{6}}\right)$ and $\left(\frac{2}{\sqrt{6}}, \frac{-1}{\sqrt{6}}, \frac{-1}{\sqrt{6}}\right)$ are vectors of unit length orthogonal to **a** and **b**.

Exercises

1. Exercise 1, page **K**271.
2. Exercise 3, page **K**272.
3. Exercise 4, page **K**272.
4. Exercise 5, page **K**272.
5. Exercise 8, page **K**272.
6. Exercise 11, page **K**272.

Solutions

1. We want to verify that $\|\mathbf{x}+\mathbf{y}\|^2 = \|\mathbf{x}\|^2 + \|\mathbf{y}\|^2$ for $\mathbf{x} = \sin$, $\mathbf{y} = \cos$.

$$\|\mathbf{x}\|^2 = \mathbf{x}\cdot\mathbf{x} = \int_{-\pi}^{\pi} \sin^2 x\, dx = \pi$$

$$\|\mathbf{y}\|^2 = \mathbf{y}\cdot\mathbf{y} = \int_{-\pi}^{\pi} \cos^2 x\, dx = \pi$$

$$\begin{aligned}\|\mathbf{x}+\mathbf{y}\|^2 &= (\mathbf{x}+\mathbf{y})\cdot(\mathbf{x}+\mathbf{y}) \\ &= \int_{-\pi}^{\pi} (\cos x + \sin x)^2\, dx \\ &= \int_{-\pi}^{\pi} (\cos^2 x + 2\cos x \sin x + \sin^2 x)\, dx \\ &= \int_{-\pi}^{\pi} dx + 2\int_{-\pi}^{\pi} \cos x \sin x\, dx = 2\pi.\end{aligned}$$

$\left(\text{See Section III. 5.1 of } \mathbf{TI} \text{ for } \int_{-\pi}^{\pi} \cos x \sin x\, dx.\right)$

Hence $\|\mathbf{x}+\mathbf{y}\|^2 = \|\mathbf{x}\|^2 + \|\mathbf{y}\|^2$ is verified.

2. We write $\mathbf{1}$, $\mathbf{x}$, $\mathbf{x}^2 - \frac{1}{3}\mathbf{1}$ for the functions $x \longmapsto 1$, $x \longmapsto x$, $x \longmapsto x^2 - \frac{1}{3}$; then we have

$$\mathbf{1}\cdot\mathbf{x} = \int_{-1}^{1} x\, dx = \left[\frac{x^2}{2}\right]_{-1}^{1} = 0$$

$$\mathbf{1}\cdot\left(\mathbf{x}^2 - \frac{1}{3}\mathbf{1}\right) = \int_{-1}^{1}\left(x^2 - \frac{1}{3}\right) dx = \left[\frac{x^3}{3} - \frac{x}{3}\right]_{-1}^{1} = 0$$

$$\begin{aligned}\mathbf{x}\cdot\left(\mathbf{x}^2 - \frac{1}{3}\mathbf{1}\right) &= \int_{-1}^{1} x\left(x^2 - \frac{1}{3}\right) dx \\ &= \left[\frac{x^4}{4} - \frac{x^2}{6}\right]_{-1}^{1} = 0,\end{aligned}$$

and hence the functions are mutually orthogonal.

3. The length of $\dfrac{\mathbf{x}}{\|\mathbf{x}\|}$ is

$$\left\|\frac{\mathbf{x}}{\|\mathbf{x}\|}\right\| = \sqrt{\frac{\mathbf{x}\cdot\mathbf{x}}{\|\mathbf{x}\|\,\|\mathbf{x}\|}} = \frac{1}{\|\mathbf{x}\|}\sqrt{\mathbf{x}\cdot\mathbf{x}} = 1$$

4. To make the set of vectors of **K**'s Exercise 3 orthonormal, we only need to make each vector normal (i.e. of length 1) as they are already orthogonal. We have seen that, if $\mathbf{x}$ is non-zero, $\mathbf{x}/\|\mathbf{x}\|$ is of unit length. Hence

$$\left\{\frac{\mathbf{1}}{\|\mathbf{1}\|}, \frac{\mathbf{x}}{\|\mathbf{x}\|}, \frac{\mathbf{x}^2 - \frac{1}{3}\mathbf{1}}{\|\mathbf{x}^2 - \frac{1}{3}\mathbf{1}\|}\right\}$$

is an orthonormal set, where

$$\|\mathbf{1}\| = \sqrt{\mathbf{1}\cdot\mathbf{1}} = \left(\int_{-1}^{1} 1\, dx\right)^{1/2} = 2^{1/2}$$

$$\|\mathbf{x}\| = \sqrt{\mathbf{x}\cdot\mathbf{x}} = \left(\int_{-1}^{1} x^2\, dx\right)^{1/2} = \left(\frac{2}{3}\right)^{1/2}$$

$$\begin{aligned}\|\mathbf{x}^2 - \tfrac{1}{3}\mathbf{1}\| &= \sqrt{(x^2 - \tfrac{1}{3}\mathbf{1})\cdot(x^2 - \tfrac{1}{3}\mathbf{1})} \\ &= \left(\int_{-1}^{1} x^4 - \tfrac{2}{3}x^2 + \tfrac{1}{9}\, dx\right)^{1/2} \\ &= \tfrac{1}{3}\sqrt{\tfrac{8}{5}}\end{aligned}$$

Thus the orthonormal set is

$$\left\{\frac{1}{\sqrt{2}}\mathbf{1}, \sqrt{\frac{3}{2}}\mathbf{x}, 3\sqrt{\frac{5}{8}}\left(\mathbf{x}^2 - \frac{1}{3}\mathbf{1}\right)\right\}.$$

5. If the vector $\mathbf{z} = (a, b, c)$ is of unit length and orthogonal to $\mathbf{x}$ and $\mathbf{y}$ we have

$$\mathbf{z}\cdot\mathbf{z} = 1, \mathbf{z}\cdot\mathbf{x} = 0, \mathbf{z}\cdot\mathbf{y} = 0;$$

i.e.
$$a^2 + b^2 + c^2 = 1,$$
$$a - b = 0,$$
$$2a + b - c = 0.$$

Reducing the last two equations to Hermite normal form gives

$$a - \tfrac{1}{3}c = 0$$
$$b - \tfrac{1}{3}c = 0$$

so that $\mathbf{z}$ must be of the form $(\frac{1}{3}c, \frac{1}{3}c, c)$. Substituting this in the first equation gives

$$\tfrac{1}{9}c^2 + \tfrac{1}{9}c^2 + c^2 = 1,$$

i.e. $c = \pm\dfrac{3}{\sqrt{11}}$

Thus we have the two solutions $\mathbf{z} = \pm\dfrac{1}{\sqrt{11}}(1, 1, 3)$.

6. Let the required combination be

$$\mathbf{f}: x \longmapsto ae^x + be^{-x} \qquad (x \in [0, 1])$$

where a, b are real constants.

For $\mathbf{f}$ and $\mathbf{exp}$ to be orthogonal in $\mathcal{C}[0, 1]$, we require

$$0 = \mathbf{f}\cdot\mathbf{exp} = \int_0^1 e^x(ae^x + be^{-x})\,dx$$
$$= \int_0^1 (ae^{2x} + b)\,dx = \left[\frac{ae^{2x}}{2} + bx\right]_0^1$$
$$= \frac{a}{2}(e^2 - 1) + b$$

so that $b = -\dfrac{a}{2}(e^2 - 1)$.

Hence the required combination is of the form

$$x \longmapsto ae^x - \frac{a}{2}(e^2 - 1)e^{-x}.$$

Note that the combination with $a = 0$ is acceptable.

16.3.2 Linear Independence of Orthogonal Sets

READ from line —9 on page **K**270 *to the end of Section 7-3 on page* **K**271.

Note

line 1, page **K**271 Note that $\mathcal{S}$ may be an infinite set.

Exercise

Show that

$$\{\mathbf{x}_1 = (1, -2, 2), \mathbf{x}_2 = (-2, 1, 2), \mathbf{x}_3 = (2, 2, 1)\}$$

forms a basis for R^3.

Solution

By Corollary 7-1 (page **K**271) we can prove a set is a basis for V by showing that it is an orthogonal set and that the number of vectors in the set equals the dimension of V. Here R^3 is 3-dimensional and we have 3 vectors in the set. To show orthogonality:

$$\mathbf{x}_1 \cdot \mathbf{x}_2 = -2 - 2 + 4 = 0.$$
$$\mathbf{x}_2 \cdot \mathbf{x}_3 = -4 + 2 + 2 = 0.$$
$$\mathbf{x}_3 \cdot \mathbf{x}_1 = 2 - 4 + 2 = 0.$$

Thus $\{\mathbf{x}_1, \mathbf{x}_2, \mathbf{x}_3\}$ is a basis.

We see that it is sometimes possible to simplify the process of determining that a set is linearly independent by introducing an inner product. Note, however, that not all linearly independent sets are orthogonal.

16.3.3 Summary of Section 16.3

In this section we defined the terms

orthogonal	(page **K**268)	* * *
orthogonal set	(page **K**268)	* * *
orthonormal set	(page **K**268)	* * *
trigonometric polynomial	(page **K**269)	* *
normalized	(page **K**269)	* * *

Theorems

1. (**7-2**, page **K**268) (Pythagoras)
Two vectors **x** and **y** in a Euclidean space are orthogonal if and only if * * *

$$\|\mathbf{x} + \mathbf{y}\|^2 = \|\mathbf{x}\|^2 + \|\mathbf{y}\|^2.$$

2. (**7-3**, page **K**270)
Every orthogonal set of vectors in a Euclidean space $\mathcal{V}$ is linearly independent. * * *

Technique

Given two vectors in a Euclidean space, find out whether or not they are orthogonal. * * *

16.4 ORTHOGONALIZATION

16.4.1 Gram-Schmidt Process

READ from page **K**273 *to line 4 of page* **K**277.

Note

line 7, page **K**273. The notation $\mathcal{S}(\mathcal{X})$ means the same as $\langle\mathcal{X}\rangle$.

Exercises

1. Exercise 1(a), page **K**279.
2. Exercise 3(a), page **K**279.

Solutions

1. To orthogonalize $\{(1, 1, 0), (-1, 1, 0), (-1, 1, 1)\}$, we take $\mathbf{e}_1 = (1, 1, 0)$, and then

$$\mathbf{e}_2 = (-1, 1, 0) - \alpha(1, 1, 0)$$

such that $\mathbf{e}_2 \cdot \mathbf{e}_1 = 0$. Thus

$$0 = (-1, 1, 0) \cdot (1, 1, 0) - \alpha(1, 1, 0) \cdot (1, 1, 0)$$
$$0 = 0 - 2\alpha,$$

i.e. $\alpha = 0$, and $\mathbf{e}_2 = (-1, 1, 0)$.

For $\mathbf{e}_3$ we want

$$\mathbf{e}_3 = (-1, 1, 1) - \beta(-1, 1, 0) - \gamma(1, 1, 0)$$

such that $\mathbf{e}_3 \cdot \mathbf{e}_2 = 0 = \mathbf{e}_3 \cdot \mathbf{e}_1$

The condition

$$0 = \mathbf{e}_3 \cdot \mathbf{e}_1 = 0 - 0\beta - 2\gamma$$

implies $\gamma = 0$

and

$$0 = \mathbf{e}_3 \cdot \mathbf{e}_2 = 2 - 2\beta - 0\gamma$$

implies $\beta = 1$.

Hence $\mathbf{e}_3 = (-1, 1, 1) - (-1, 1, 0) - 0(1, 1, 0)$
$= (0, 0, 1)$.

The orthogonal basis is

$$\{(1, 1, 0), (-1, 1, 0), (0, 0, 1)\}.$$

2. We wish to orthogonalize the following set in $\mathscr{R}^4$

$$\{(1, 0, 0, 1), (-1, 0, 2, 1), (0, 1, 2, 0), (0, 0, -1, 1)\}$$

Take $\mathbf{e}_1 = (1, 0, 0, 1)$,

then $\mathbf{e}_2 = (-1, 0, 2, 1) - \alpha(1, 0, 0, 1)$

such that $\mathbf{e}_2 \cdot \mathbf{e}_1 = 0$.

This implies $\alpha = 0$, i.e. $(-1, 0, 2, 1)$ is already orthogonal to $\mathbf{e}_1$.

Hence $\mathbf{e}_2 = (-1, 0, 2, 1)$.

For $\mathbf{e}_3$, we have

$$\mathbf{e}_3 = (0, 1, 2, 0) - \beta(-1, 0, 2, 1) - \gamma(1, 0, 0, 1)$$

such that $\mathbf{e}_3 \cdot \mathbf{e}_1 = \mathbf{e}_3 \cdot \mathbf{e}_2 = 0$.

The condition $\mathbf{e}_3 \cdot \mathbf{e}_1 = 0$ implies $\gamma = 0$ (as $\mathbf{e}_1$ is already orthogonal to the other two vectors).

The condition $\mathbf{e}_3 \cdot \mathbf{e}_2 = 0$ gives

$$0 = 4 - 6\beta,$$

i.e. $\beta = \frac{2}{3}$.

To avoid fractions we compute $3\mathbf{e}_3$.

$$\begin{aligned}3\mathbf{e}_3 &= 3(0, 1, 2, 0) - 2(-1, 0, 2, 1)\\ &= (2, 3, 2, -2).\end{aligned}$$

i.e. $\mathbf{e}_3 = \frac{1}{3}(2, 3, 2, -2)$.

For $\mathbf{e}_4$, we have

$$\begin{aligned}\mathbf{e}_4 = (0, 0, -1, 1) &- \delta(2, 3, 2, -2)\\ &-\varepsilon(-1, 0, 2, 1) - \zeta(1, 0, 0, 1)\end{aligned}$$

such that $\mathbf{e}_4 \cdot \mathbf{e}_3 = \mathbf{e}_4 \cdot \mathbf{e}_2 = \mathbf{e}_4 \cdot \mathbf{e}_1 = 0$.

These conditions give

$$\begin{aligned}&0 = -4 - 21\delta, \text{ i.e. } \delta = -\tfrac{4}{21},\\ &0 = -1 - 6\varepsilon, \text{ i.e. } \varepsilon = -\tfrac{1}{6},\\ &0 = 1 - 2\zeta, \text{ i.e. } \zeta = \tfrac{1}{2}.\end{aligned}$$

To avoid fractions we compute $42\mathbf{e}_4$.

$$\begin{aligned}42\mathbf{e}_4 &= 42(0, 0, -1, 1) + 8(2, 3, 2, -2) + 7(-1, 0, 2, 1)\\ &\quad -21(1, 0, 0, 1)\\ &= 12(-1, 2, -1, 1)\end{aligned}$$

i.e. $\mathbf{e}_4 = \frac{2}{7}(-1, 2, -1, 1)$.

Hence the orthogonalization process gives

$$\{(1, 0, 0, 1), (-1, 0, 2, 1), \tfrac{1}{3}(2, 3, 2, -2), \tfrac{2}{7}(-1, 2, -1, 1)\}$$

16.4.2 Orthogonal Polynomials

READ from line 5, page **K**277 *to line 6, page* **K**278.

Note

line 12, page **K**277 It is common practice in Leibniz notation to omit the integrand $f(x)$ when $f(x) = 1$.

Exercise

Continue Example 3 of the reading passage to find the polynomials $\mathbf{e}_4$ and $\mathbf{e}_5$.

Solution

Continuing the process, we require

$$\mathbf{e}_4 = \mathbf{x}^3 - \alpha_3 \mathbf{e}_3 - \alpha_2 \mathbf{e}_2 - \alpha_1 \mathbf{e}_1$$

such that

$$\mathbf{e}_4 \cdot \mathbf{e}_3 = \mathbf{e}_4 \cdot \mathbf{e}_2 = \mathbf{e}_4 \cdot \mathbf{e}_1 = 0.$$

These equations give

$$\left.\begin{aligned}\mathbf{x}^3 \cdot \mathbf{e}_3 &= \alpha_3 \mathbf{e}_3 \cdot \mathbf{e}_3\\ \mathbf{x}^3 \cdot \mathbf{e}_2 &= \alpha_2 \mathbf{e}_2 \cdot \mathbf{e}_2\\ \mathbf{x}^3 \cdot \mathbf{e}_1 &= \alpha_1 \mathbf{e}_1 \cdot \mathbf{e}_1\end{aligned}\right\} \qquad (1)$$

We know that (see page **K**277)

$$\mathbf{e}_1 \cdot \mathbf{e}_1 = 2, \qquad \mathbf{e}_2 \cdot \mathbf{e}_2 = \tfrac{2}{3}, \qquad \mathbf{e}_3 \cdot \mathbf{e}_3 = \tfrac{8}{45}.$$

Also $\mathbf{x}^3 \cdot \mathbf{e}_3 = \int_{-1}^{1} x^3(x^2 - \tfrac{1}{3})\, dx = 0,$

$$\mathbf{x}^3 \cdot \mathbf{e}_1 = \int_{-1}^{1} x^3\, dx = 0,$$

$$\mathbf{x}^3 \cdot \mathbf{e}_2 = \int_{-1}^{1} x^4\, dx = \tfrac{2}{5}.$$

Thus Equations (1) give

$$\alpha_3 = 0, \alpha_2 = \tfrac{3}{5}, \alpha_1 = 0$$

and so

$$\mathbf{e}_4 = \mathbf{x}^3 - \tfrac{3}{5}\mathbf{x}.$$

For $\mathbf{e}_5$ we require

$$\mathbf{e}_5 = \mathbf{x}^4 - \beta_4 \mathbf{e}_4 - \beta_3 \mathbf{e}_3 - \beta_2 \mathbf{e}_2 - \beta_1 \mathbf{e}_1$$

such that

$$\mathbf{e}_5 \cdot \mathbf{e}_4 = \mathbf{e}_5 \cdot \mathbf{e}_3 = \mathbf{e}_5 \cdot \mathbf{e}_2 = \mathbf{e}_5 \cdot \mathbf{e}_1 = 0.$$

This gives

$$\left.\begin{aligned} \mathbf{x}^4 \cdot \mathbf{e}_4 &= \beta_4 \mathbf{e}_4 \cdot \mathbf{e}_4 \\ \mathbf{x}^4 \cdot \mathbf{e}_3 &= \beta_3 \mathbf{e}_3 \cdot \mathbf{e}_3 \\ \mathbf{x}^4 \cdot \mathbf{e}_2 &= \beta_2 \mathbf{e}_2 \cdot \mathbf{e}_2 \\ \mathbf{x}^4 \cdot \mathbf{e}_1 &= \beta_1 \mathbf{e}_1 \cdot \mathbf{e}_1 \end{aligned}\right\} \qquad (2)$$

Also

$$\mathbf{x}^4 \cdot \mathbf{e}_4 = \int_{-1}^{1} x^4\left(x^3 - \frac{3}{5}x\right) dx = 0$$

$$\mathbf{x}^4 \cdot \mathbf{e}_2 = \int_{-1}^{1} x^4(x)\, dx = 0$$

Hence $\beta_4 = \beta_2 = 0$.

The remaining inner products we require are

$$\mathbf{x}^4 \cdot \mathbf{e}_3 = \int_{-1}^{1} x^4\left(x^2 - \frac{1}{3}\right) dx$$

$$= \tfrac{16}{105}$$

and

$$\mathbf{x}^4 \cdot \mathbf{e}_1 = \int_{-1}^{1} x^4\, dx = \tfrac{2}{5}.$$

Thus Equations (2) give

$$\beta_3 = \tfrac{6}{7} \quad \text{and} \quad \beta_1 = \tfrac{1}{5}$$

and, so

$$\mathbf{e}_5 = x^4 - \tfrac{6}{7}x^2 + \tfrac{3}{35}.$$

Exercise

Exercise 17, page **K**280.

Solution

(a) $P_0(x) = \frac{0!}{2^0(0!)^2} x^0 = 1$

$P_1(x) = \frac{2!}{2(1!)^2} x = x$

$P_2(x) = \frac{4!}{2^2(2!)^2}\left(x^2 - \frac{2 \times 1}{2 \times 3}\right) = \frac{3}{2}(x^2 - \frac{1}{3})$

$P_3(x) = \frac{6!}{2^3(3!)^2}\left(x^3 - \frac{3 \times 2}{2 \times 5} x\right) = \frac{5}{2}(x^3 - \frac{3}{5}x)$

$P_4(x) = \frac{8!}{2^4(4!)^2}\left(x^4 - \frac{4 \times 3}{2 \times 7} x^2 + \frac{4 \times 3 \times 2 \times 1}{2 \times 4 \times 7 \times 5}\right)$

$= \frac{35}{8}(x^4 - \frac{6}{7}x^2 + \frac{3}{35})$

(b) The answers are given on page **K**747. We illustrate the method for part (i) only.

$P_0, \ldots, P_4$ are mutually orthogonal since they are constant multiples of the orthogonal sequence which was the subject of the previous exercise.

$$3x^2 - 2x + 1 = \alpha_0 P_0(x) + \alpha_1 P_1(x) + \alpha_2 P_2(x),$$

since $P_3(x)$, $P_4(x)$, ... involve higher powers of x than x^2. Then, if we take the inner product of both sides with $P_0(x)$, we have

$$\int_{-1}^{1} (3x^2 - 2x + 1) P_0(x)\, dx = \alpha_0 \int_{-1}^{1} (P_0(x))^2\, dx$$

since P_0, P_1 and P_2 are mutually orthogonal.

Hence $\alpha_0 = 2$.

Similarly, by taking inner products with $P_1(x)$ and $P_2(x)$, α_1 and α_2 can be found.

16.4.3 Orthonormal Basis

READ page **K**278 *from line* 7 *to the end of Section* 7-4.

Note

lines −2 and −1, page **K**278   We met Exercise 10 in sub-section 16.1.2.

Example

We have shown (see the exercise in sub-section 16.3.2) that

$$\{\mathbf{x}_1 = (1, -2, 2),\ \mathbf{x}_2 = (-2, 1, 2),\ \mathbf{x}_3 = (2, 2, 1)\}$$

is an *orthogonal* basis for R^3. But notice that $\|\mathbf{x}_1\| = \|\mathbf{x}_2\| = \|\mathbf{x}_3\| = 3$

An *orthonormal* basis in R^3 would be $\{\frac{1}{3}\mathbf{x}_1, \frac{1}{3}\mathbf{x}_2, \frac{1}{3}\mathbf{x}_3\}$. Call it

$$\{\mathbf{e}_1 = \tfrac{1}{3}\mathbf{x}_1,\ \mathbf{e}_2 = \tfrac{1}{3}\mathbf{x}_2,\ \mathbf{e}_3 = \tfrac{1}{3}\mathbf{x}_3\}.$$

Suppose we have two vectors

$\mathbf{a} = \mathbf{e}_1 - \frac{1}{2}\mathbf{e}_2 + 3\mathbf{e}_3$

$\mathbf{b} = \frac{1}{2}\mathbf{e}_1 + 3\mathbf{e}_2 - \frac{1}{3}\mathbf{e}_3.$

We can now use Equation (7-47) to give $\mathbf{a} \cdot \mathbf{b}$.

$\mathbf{a} \cdot \mathbf{b} = \frac{1}{2} - \frac{3}{2} - 1 = -2.$

Exercises

1. Exercise 9, page **K**279.
2. Exercise 10, page **K**280.
3. Exercise 11, page **K**280.
4. Exercise 14, page **K**280.

Solutions

1. (a) If $\mathbf{e}_{n+1} = \mathbf{0}$ then

$$\mathbf{x}_{n+1} = \alpha_1 \mathbf{e}_1 + \cdots + \alpha_n \mathbf{e}_n$$

This gives $\mathbf{x}_{n+1} \in \mathcal{S}(\mathbf{e}_1, \ldots, \mathbf{e}_n)$ and hence $\mathbf{x}_{n+1} \in \mathcal{S}(\mathbf{x}_1, \ldots, \mathbf{x}_n)$. But $\{\mathbf{x}_1, \ldots, \mathbf{x}_n, \mathbf{x}_{n+1}\}$ is linearly independent and hence $\mathbf{x}_{n+1} \notin \mathcal{S}(\mathbf{x}_1, \ldots, \mathbf{x}_n)$. This contradiction proves that the assertion $\mathbf{e}_{n+1} = \mathbf{0}$ is false and hence $\mathbf{e}_{n+1} \neq \mathbf{0}$.

(b) $\mathcal{S}(\mathbf{e}_1, \ldots, \mathbf{e}_n) = \mathcal{S}(\mathbf{x}_1, \ldots, \mathbf{x}_n) \subseteq \mathcal{S}(\mathbf{x}_1, \ldots, \mathbf{x}_{n+1})$ and $\mathbf{e}_{n+1} = \mathbf{x}_{n+1} - \alpha_1 \mathbf{e}_1 - \cdots - \alpha_n \mathbf{e}_n$ implies

$$\mathbf{e}_{n+1} \in \mathcal{S}(\mathbf{e}_1, \ldots, \mathbf{e}_n, \mathbf{x}_{n+1}) = \mathcal{S}(\mathbf{x}_1, \ldots, \mathbf{x}_n, \mathbf{x}_{n+1}).$$

Hence,

$$\mathcal{S}(\mathbf{e}_1, \ldots, \mathbf{e}_n, \mathbf{e}_{n+1}) \subseteq \mathcal{S}(\mathbf{x}_1, \ldots, \mathbf{x}_n, \mathbf{x}_{n+1}).$$

Also

$$\mathcal{S}(\mathbf{x}_1, \ldots, \mathbf{x}_n) = \mathcal{S}(\mathbf{e}_1, \ldots, \mathbf{e}_n) \subseteq \mathcal{S}(\mathbf{e}_1, \ldots, \mathbf{e}_n, \mathbf{e}_{n+1})$$

and

$$\mathbf{x}_{n+1} = \mathbf{e}_{n+1} + \alpha_1 \mathbf{e}_1 + \cdots + \alpha_n \mathbf{e}_n$$
$$\mathbf{x}_{n+1} \in \mathcal{S}(\mathbf{e}_1, \ldots, \mathbf{e}_{n+1})$$

Hence,

$$\mathcal{S}(\mathbf{x}_1, \ldots, \mathbf{x}_n, \mathbf{x}_{n+1}) \subseteq \mathcal{S}(\mathbf{e}_1, \ldots, \mathbf{e}_n, \mathbf{e}_{n+1}).$$

The two inclusions combined imply

$$\mathcal{S}(\mathbf{x}_1, \ldots, \mathbf{x}_{n+1}) = \mathcal{S}(\mathbf{e}_1, \ldots, \mathbf{e}_{n+1}).$$

2. Since $\mathbf{x} = \sum_{i=1}^{n} \alpha_i \mathbf{e}_i$ and $\mathbf{y} = \sum_{j=1}^{n} \beta_j \mathbf{e}_j$,

$$\mathbf{x} \cdot \mathbf{y} = \left(\sum_{i=1}^{n} \alpha_i \mathbf{e}_i\right) \cdot \left(\sum_{j=1}^{n} \beta_j \mathbf{e}_j\right)$$
$$= \sum_{i=1}^{n} \sum_{j=1}^{n} \alpha_i \beta_j (\mathbf{e}_i \cdot \mathbf{e}_j), \text{ by bilinearity.}$$

Now, because $\{\mathbf{e}_1, \ldots, \mathbf{e}_n\}$ is given as an orthonormal basis $\mathbf{e}_i \cdot \mathbf{e}_j = \delta_{ij}$, where $\delta_{ij} = 1$ if $i = j$ and 0 otherwise.

Hence

$$\mathbf{x} \cdot \mathbf{y} = \sum_{i=1}^{n} \sum_{j=1}^{n} \alpha_i \beta_j \delta_{ij}$$

and the only non-zero terms will occur when $i = j$.

Hence

$$\mathbf{x} \cdot \mathbf{y} = \sum_{i=1}^{n} \alpha_i \beta_i.$$

3. (a) $\mathbf{x} \cdot \mathbf{e}_1 = (\alpha_1 \mathbf{e}_1 + \alpha_2 \mathbf{e}_2 + \cdots + \alpha_n \mathbf{e}_n) \cdot \mathbf{e}_1$

$$= \alpha_1 (\mathbf{e}_1 \cdot \mathbf{e}_1) + \cdots + \alpha_n (\mathbf{e}_n \cdot \mathbf{e}_1)$$
$$= \alpha_1 (\mathbf{e}_1 \cdot \mathbf{e}_1),$$

since $\mathbf{e}_1 \cdot \mathbf{e}_j = 0$ for $j \neq 1$.

Hence, $\alpha_1 = (\mathbf{x} \cdot \mathbf{e}_1)/(\mathbf{e}_1 \cdot \mathbf{e}_1)$.

Similarly, $\alpha_i = (\mathbf{x} \cdot \mathbf{e}_i)/(\mathbf{e}_i \cdot \mathbf{e}_i)$, for $i = 2, 3, \ldots, n$.

(b) If $\mathbf{x} = \sum_{i=1}^{n} \alpha_i \mathbf{e}_i$ and $\mathbf{y} = \sum_{i=1}^{n} \beta_i \mathbf{e}_i$,

$$\mathbf{x} \cdot \mathbf{y} = \sum_{i,j} \alpha_i \beta_j (\mathbf{e}_i \cdot \mathbf{e}_j)$$

But since the basis is orthogonal $\mathbf{e}_i \cdot \mathbf{e}_j = 0\ (i \neq j)$, hence

$$\mathbf{x} \cdot \mathbf{y} = \sum_{i=1}^{n} \alpha_i \beta_i (\mathbf{e}_i \cdot \mathbf{e}_i)$$

$$= \sum_{i=1}^{n} \frac{(\mathbf{x} \cdot \mathbf{e}_i)}{(\mathbf{e}_i \cdot \mathbf{e}_i)} \frac{(\mathbf{y} \cdot \mathbf{e}_i)}{(\mathbf{e}_i \cdot \mathbf{e}_i)} (\mathbf{e}_i \cdot \mathbf{e}_i), \text{ by part (a).}$$

$$= \sum_{i=1}^{n} (\mathbf{x} \cdot \mathbf{e}_i)(\mathbf{y} \cdot \mathbf{e}_i)/(\mathbf{e}_i \cdot \mathbf{e}_i)$$

4. (a) In $\mathcal{C}[-\pi, \pi]$

$$\mathbf{f} \cdot \mathbf{g} = \int_{-\pi}^{\pi} f(x) g(x)\, dx.$$

To orthogonalize $\{1, \sin, \sin^2\}$, we take $\mathbf{e}_1 = 1$ and look for

$$\mathbf{e}_2 = \sin - \alpha 1$$

such that

$$\mathbf{e}_2 \cdot \mathbf{e}_1 = 0.$$

$$\mathbf{e}_2 \cdot \mathbf{e}_1 = \int_{-\pi}^{\pi} (\sin x - \alpha)\, dx$$

$$= [-\cos x - \alpha x]_{-\pi}^{\pi} = 0 - 2\alpha\pi.$$

Hence $\alpha = 0$, and so

$$\mathbf{e}_2 = \sin.$$

For $\mathbf{e}_3$, we want $\mathbf{e}_3 = \sin^2 - \alpha_1 \sin - \alpha_2$ such that $\mathbf{e}_3 \cdot \mathbf{e}_2 = \mathbf{e}_3 \cdot \mathbf{e}_1 = 0$.

$$0 = \mathbf{e}_3 \cdot \mathbf{e}_1 = \int_{-\pi}^{\pi} (\sin^2 x - \alpha_1 \sin x - \alpha_2)\, dx$$

$$= \int_{-\pi}^{\pi} \left(\left(\frac{1 - \cos 2x}{2} \right) - \alpha_1 \sin x - \alpha_2 \right) dx$$

$$= [(\tfrac{1}{2} - \alpha_2)x]_{-\pi}^{\pi} = (\tfrac{1}{2} - \alpha_2)2\pi.$$

Hence $\alpha_2 = \frac{1}{2}$

$$0 = \mathbf{e}_3 \cdot \mathbf{e}_2 = \int_{-\pi}^{\pi} (\sin^3 x - \alpha_1 \sin^2 x - \alpha_2 \sin x)\, dx$$

$$= \int_{-\pi}^{\pi} \left(\sin^3 x - \alpha_1 \left(\frac{1 - \cos 2x}{2} \right) - \alpha_2 \sin x \right) dx$$

$$= \left[\frac{-\alpha_1 x}{2} \right]_{-\pi}^{\pi} = -\alpha_1 \pi.$$

For $\int_{-\pi}^{\pi} \sin^3 x\, dx$, see Section III. 4.2 of **TI**.)

Hence $\alpha_1 = 0$ and $\mathbf{e}_3 = \sin^2 - \frac{1}{2}$.

Hence the orthogonalized set is $\{1, \sin, \sin^2 - \frac{1}{2}\}$.

(b) We already know that $\sin^2 - \frac{1}{2}$ is orthogonal to $\langle 1, \sin \rangle$, hence all we need to do is normalize it. The vector we want is

$$\frac{\sin^2 - \frac{1}{2}}{\|\sin^2 - \frac{1}{2}\|}.$$

Now $\|\sin^2 - \frac{1}{2}\|^2 = \int_{-\pi}^{\pi} (\sin^2 x - \frac{1}{2})^2\, dx.$

But $(\sin^2 x - \frac{1}{2}) = -\frac{1}{2}\cos 2x.$

The integral becomes $\int_{-\pi}^{\pi} \frac{1}{4}\cos^2 2x\, dx$

$$= \frac{1}{8}\int_{-\pi}^{\pi} (\cos 4x + 1)\, dx$$

$$= \left[\frac{x}{8}\right]_{-\pi}^{\pi} = \frac{\pi}{4}.$$

Hence, the vector we require is $\sqrt{\frac{4}{\pi}}\,(\sin^2 - \frac{1}{2}).$

16.4.4 Summary of Section 16.4

In this section we define the term

Gram-Schmidt orthogonalization process (page **K**277) * * *

Theorem

(**7-4**, page **K**276)

Let $\mathbf{x}_1, \mathbf{x}_2, \ldots$ be a (finite or infinite) set of linearly independent vectors in a Euclidean space $\mathcal{V}$ such that for each integer n, $\mathcal{S}(\mathbf{e}_1, \ldots, \mathbf{e}_n) = \mathcal{S}(\mathbf{x}_1, \ldots, \mathbf{x}_n)$. Moreover, the $\mathbf{e}_n$ can be chosen according to the rules * * *

$$\mathbf{e}_1 = \mathbf{x}_1$$

and

$$\mathbf{e}_{n+1} = \mathbf{x}_{n+1} - \alpha_1 \mathbf{e}_1 - \cdots - \alpha_n \mathbf{e}_n,$$

where

$$\alpha_1 = \frac{\mathbf{x}_{n+1} \cdot \mathbf{e}_1}{\mathbf{e}_1 \cdot \mathbf{e}_1},\ \alpha_2 = \frac{\mathbf{x}_{n+1} \cdot \mathbf{e}_2}{\mathbf{e}_2 \cdot \mathbf{e}_2},\ \ldots,\ \alpha_n = \frac{\mathbf{x}_{n+1} \cdot \mathbf{e}_n}{\mathbf{e}_n \cdot \mathbf{e}_n}.$$

Technique

Given a set of linearly independent vectors, use the Gram-Schmidt orthogonalization process to construct an orthogonal set spanning the same subspace. * * *

16.5 SUMMARY OF THE UNIT

A Euclidean space is a mathematical structure consisting of a real vector space together with a particular type of bilinear form called an inner product.

Starting from this definition developed in the first section we went on to define length, angle and distance in a Euclidean space in the second section. The third section looked at orthogonal vectors in a Euclidean space and the fourth section introduced a technique for orthogonalizing any linearly independent set of vectors—the Gram-Schmidt process.

Definitions

inner product	(page **K**256)	* * *
Euclidean space	(page **K**256)	* * *
Euclidean *n*-space	(page **K**257)	* * *
weight function	(page **K**258)	* * *
length (norm)	(page **K**261)	* * *
angle	(page **K**263)	* *
distance	(page **K**264)	* * *
orthogonal	(page **K**268)	* * *
orthogonal set	(page **K**268)	* * *
orthonormal set	(page **K**268)	* * *
trigonometric polynomial	(page **K**269)	* *
normalized	(page **K**269)	* * *
Gram-Schmidt orthogonalization process	(page **K**277)	* * *

The most important inner products for our purposes are

$$\mathbf{x} \cdot \mathbf{y} = \sum_{i=1}^{n} x_i y_i \quad \text{in} \quad R^n \qquad \text{(page } \mathbf{K}257\text{)}$$ * * *

$$\mathbf{f} \cdot \mathbf{g} = \int_a^b f(x)g(x)r(x)\,dx \quad \text{in} \quad C[a, b] \qquad \text{(page } \mathbf{K}258\text{)}$$ * * *

where $r(x)$ is non-negative with only a finite number of zeros in $C[a, b]$.

Theorems

1. (**7-1**, page **K**262) (Schwarz inequality)
If $\mathbf{x}$ and $\mathbf{y}$ are any two vectors in a Euclidean space, then * * *

$$(\mathbf{x} \cdot \mathbf{y})^2 \leqslant (\mathbf{x} \cdot \mathbf{x})(\mathbf{y} \cdot \mathbf{y}).$$

2. (**Lemma 7-1**, page **K**264) (triangle inequality)
If $\mathbf{x}$ and $\mathbf{y}$ are arbitrary vectors in a Euclidean space, then * * *

$$\|\mathbf{x} + \mathbf{y}\| \leqslant \|\mathbf{x}\| + \|\mathbf{y}\|.$$

3. (**7-2**, page **K**268) (Pythagoras)
Two vectors $\mathbf{x}$ and $\mathbf{y}$ in a Euclidean space are orthogonal if and only if * * *

$$\|\mathbf{x} + \mathbf{y}\|^2 = \|\mathbf{x}\|^2 + \|\mathbf{y}\|^2.$$

4. (**7-3**, page **K**270)
Every orthogonal set of vectors in a Euclidean space $\mathcal{V}$ is linearly independent. * * *

5. (**7-4**, page **K**276)
Let $\mathbf{x}_1, \mathbf{x}_2, \ldots$ be a (finite or infinite) set of linearly independent vectors in a Euclidean space $\mathcal{V}$ such that for each integer n, $\mathcal{S}(\mathbf{e}_1, \ldots, \mathbf{e}_n) = \mathcal{S}(\mathbf{x}_1, \ldots, \mathbf{x}_n)$. Moreover, the $\mathbf{e}_n$ can be chosen according to the rules * * *

$$\mathbf{e}_1 = \mathbf{x}_1$$

and

$$\mathbf{e}_{n+1} = \mathbf{x}_{n+1} - \alpha_1 \mathbf{e}_1 - \cdots - \alpha_n \mathbf{e}_n,$$

where

$$\alpha_1 = \frac{\mathbf{x}_{n+1} \cdot \mathbf{e}_1}{\mathbf{e}_1 \cdot \mathbf{e}_1},\ \alpha_2 = \frac{\mathbf{x}_{n+1} \cdot \mathbf{e}_2}{\mathbf{e}_2 \cdot \mathbf{e}_2},\ \ldots,\ \alpha_n = \frac{\mathbf{x}_{n+1} \cdot \mathbf{e}_n}{\mathbf{e}_n \cdot \mathbf{e}_n}$$

Techniques

1. Given two vectors in R^n or two functions in $C[a, b]$, find their inner product. * * *

2. Given two vectors in a Euclidean space, find their lengths and the angle and distance between them. * * *

3. Given two vectors in a Euclidean space, find out whether or not they are orthogonal. * * *

4. Given a set of linearly independent vectors, use the Gram-Schmidt process to construct an orthogonal set spanning the same subspace. * * *

Notation

$\|\mathbf{x}\|$	(page **K**261)
$d(\mathbf{x}, \mathbf{y})$	(page **K**264)

16.6 SELF-ASSESSMENT

Self-assessment Test

This Self-assessment Test is designed to help you test your understanding of the unit. It can also be used, together with the summary of the unit, for revision. The answers to these questions will be found on the next non-facing page. We suggest that you complete the whole test before looking at the answers.

1. Are the following inner products?

(i) $\mathbf{x} \cdot \mathbf{y} = x_1 y_1 + 2x_2 y_2 \quad (\mathbf{x}, \mathbf{y} \in R^2)$

(ii) $\mathbf{x} \cdot \mathbf{y} = 2x_1 y_1 + x_1 y_2 + x_2 y_1 + \frac{1}{2} x_2 y_2 \quad (\mathbf{x}, \mathbf{y} \in R^2)$

(iii) $\mathbf{f} \cdot \mathbf{g} = \int_0^{\frac{1}{2}} f(x) g(x)\, dx \quad (\mathbf{f}, \mathbf{g} \in C[0, 1])$

(iv) $\mathbf{f} \cdot \mathbf{g} = \int_0^1 (x - \frac{1}{2} + |x - \frac{1}{2}|) f(x) g(x)\, dx \quad (\mathbf{f}, \mathbf{g} \in C[0, 1])$

2. $\mathbf{x} = (2, -3, 1)$ and $\mathbf{y} = (3, 3, 2)$. Use the standard inner product for R^3 $(\mathbf{x} \cdot \mathbf{y} = x_1 y_1 + x_2 y_2 + x_3 y_3)$ to find:

(i) the length (norm) of **x**,
(ii) the angle between **x** and **y**,
(iii) the distance between the two vectors **x** and **y**.

3. For $\mathbf{f}, \mathbf{g} \in C[0, 1]$ with the usual inner product (i.e. when the weight function $r(x) = 1$), what are $\|\mathbf{f}\|$ and $d(\mathbf{f}, \mathbf{g})$ in terms of the inner product?

4. Which of the following pairs of vectors in R^2 or $C[0, 1]$ (with their usual inner product respectively) are orthogonal?

(i) $(0, 1), (1, 0)$
(ii) $(1, 1), (1, -1)$
(iii) $(1, 2), (1, 1)$
(iv) $x \longmapsto x,\ x \longmapsto x^2$
(v) $x \longmapsto e^x,\ x \longmapsto x + 3e^{-x}$
(vi) $x \longmapsto e^x,\ x \longmapsto 1 + (1 - e)e^{-x}$

5. Orthogonalize the following sets of vectors in R^3 with the standard inner product:

(i) $(1, 1, 0), (1, -1, 1), (1, 2, 1)$
(ii) $(1, 0, 0), (1, 1, 0), (1, 1, 1)$

6. Find a vector which is orthogonal to $\mathbf{x} = (1, 1, 4)$ and $\mathbf{y} = (2, -2, 6)$ in R^3.

7. Orthogonalize the set of vectors $\{\mathbf{1}, \mathbf{x}, \mathbf{x}^2\}$, in the Euclidean space formed from $C[0, \infty)$ by giving it the following inner product

$$\mathbf{f} \cdot \mathbf{g} = \int_0^{\infty} e^{-x} f(x) g(x)\, dx$$

Normalize the resulting vectors.

Solutions to Self-assessment Test

1. (i) Yes: it is a real symmetric positive-definite bilinear form.
(ii) No: it is a symmetric bilinear form but it is not positive definite; for example $(1, -2) \cdot (1, -2) = 0$.
(iii) No: it is not positive-definite. For intance, if

$$\mathbf{f}: x \longmapsto 0 \qquad 0 \leqslant x < \tfrac{1}{2}$$
$$x \longmapsto \tfrac{1}{16} - (x - \tfrac{3}{4})^2 \qquad \tfrac{1}{2} \leqslant x \leqslant 1$$

then $\mathbf{f}$ is in $C[0, 1]$ and $\mathbf{f} \neq \mathbf{0}$, but $\mathbf{f} \cdot \mathbf{f} = \int_0^{\frac{1}{2}} [f(x)]^2 \, dx = 0$ because $\mathbf{f}(x) = 0$ for $0 \leqslant x \leqslant \frac{1}{2}$.

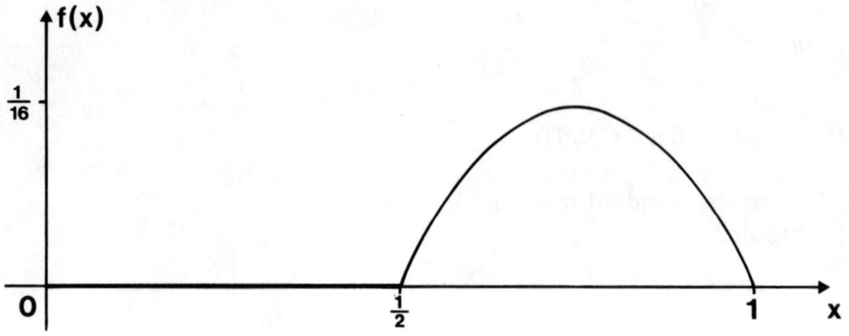

(iv) No: Again this fails to be positive definite. The weight function $r: x \longmapsto (x - \frac{1}{2} + |x - \frac{1}{2}|)$ is zero for $0 \leqslant x \leqslant \frac{1}{2}$.

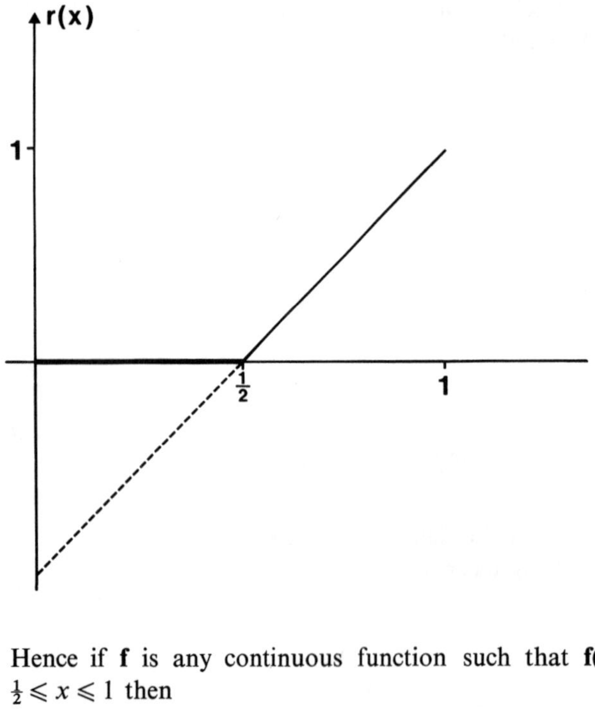

Hence if $\mathbf{f}$ is any continuous function such that $\mathbf{f}(x) = 0$ for $\frac{1}{2} \leqslant x \leqslant 1$ then

$$\mathbf{f} \cdot \mathbf{f} = \int_0^1 r(x) f(x) f(x) \, dx = 0$$

Since it is possible to find such functions $\mathbf{f}$ which are not always zero, condition (7-4), page **K**256, is not satisfied.

2. (i) $\|\mathbf{x}\| = \sqrt{(2, -3, 1) \cdot (2, -3, 1)} = \sqrt{14}$.

(ii) $\theta = \cos^{-1} \dfrac{\mathbf{x} \cdot \mathbf{y}}{\|\mathbf{x}\| \|\mathbf{y}\|} = \cos^{-1} \dfrac{-1}{\sqrt{14}\sqrt{22}} = \cos^{-1} \dfrac{-1}{\sqrt{2}\sqrt{77}}$.

(iii) $\mathbf{x} - \mathbf{y} = -(1, 6, 1)$, $d(\mathbf{x}, \mathbf{y}) = \|\mathbf{x} - \mathbf{y}\| = \sqrt{38}$.

3. $\|\mathbf{f}\| = \sqrt{\mathbf{f}\cdot\mathbf{f}} = \left(\int_0^1 [f(x)]^2\,dx\right)^{1/2}$,

$$d(\mathbf{f}, \mathbf{g}) = \|\mathbf{f}-\mathbf{g}\| = \sqrt{(\mathbf{f}-\mathbf{g})\cdot(\mathbf{f}-\mathbf{g})} = \left(\int_0^1 [f(x) - g(x)]^2\,dx\right)^{1/2}$$

4. (i) Yes: $(0, 1)\cdot(1, 0) = 0$.
(ii) Yes: $(1, 1)\cdot(1, -1) = 0$.
(iii) No: $(1, 2)\cdot(1, 1) = 3 \neq 0$.

(iv) No: $\int_0^1 x^3\,dx = \frac{1}{4} \neq 0$.

(v) No: $\int_0^1 (xe^x + 3)\,dx > 0$.

(vi) Yes: $\int_0^1 e^x + 1 - e\,dx = 0$.

5. (i) Take $\mathbf{e}_1 = (1, 1, 0)$, then $\mathbf{e}_2 = (1, -1, 1) - \alpha(1, 1, 0)$ where α is chosen such that $\mathbf{e}_1\cdot\mathbf{e}_2 = 0$, i.e. $0 = 0 - 2\alpha$, $\alpha = 0$.

Hence $\mathbf{e}_2 = (1, -1, 1)$.

Take $\mathbf{e}_3 = (1, 2, 1) - \beta(1, 1, 0) - \gamma(1, -1, 1)$ where β, γ are chosen so that $\mathbf{e}_3\cdot\mathbf{e}_1 = \mathbf{e}_3\cdot\mathbf{e}_2 = 0$.

The first condition gives $0 = 3 - 2\beta - 0$, $\quad \beta = \frac{3}{2}$.

The second condition gives $0 = -3\gamma$, $\quad \gamma = 0$.

Hence $\mathbf{e}_3 = (1, 2, 1) - \frac{3}{2}(1, 1, 0) = \frac{1}{2}(-1, 1, 2)$.

The orthogonalized set is

$$\{(1, 1, 0), (1, -1, 1), \tfrac{1}{2}(-1, 1, 2)\}.$$

(ii) Take $\mathbf{e}_1 = (1, 0, 0)$, then $\mathbf{e}_2 = (1, 1, 0) - \alpha(1, 0, 0)$ where α is chosen so that $\mathbf{e}_2\cdot\mathbf{e}_1 = 0$. This gives $\alpha = 1$ and $\mathbf{e}_2 = (0, 1, 0)$.

Take $\mathbf{e}_3 = (1, 1, 1) - \beta(1, 0, 0) - \gamma(0, 1, 0)$ where β, γ are chosen so that $\mathbf{e}_3\cdot\mathbf{e}_1 = \mathbf{e}_3\cdot\mathbf{e}_2 = 0$. This gives

$$1 - \beta = 0 \qquad 1 - \gamma = 1.$$

Hence $\beta = 1$, $\gamma = 1$ and $\mathbf{e}_3 = (0, 0, 1)$.

The orthogonalized set is then just the standard basis

$$\{(1, 0, 0), (0, 1, 0), (0, 0, 1)\}.$$

6. Let $\mathbf{z} = (a, b, c)$ be a vector orthogonal to both $\mathbf{x}$ and $\mathbf{y}$. Then

$$\mathbf{x}\cdot\mathbf{z} = 0 = a + b + 4c$$
$$\mathbf{y}\cdot\mathbf{z} = 0 = 2a - 2b + 6c$$

These equations reduce to (Hermite normal form)

$$a + 3\tfrac{1}{2}c = 0$$
$$b + \tfrac{1}{2}c = 0$$

Taking $c = -2$ we obtain a solution $\mathbf{z} = (7, 1, -2)$.

7. Take $\mathbf{e}_1 = 1$, then $\mathbf{e}_2 = x - \alpha$ where α is a constant chosen so that $\mathbf{e}_2 \cdot \mathbf{e}_1 = 0$. This condition is

$$0 = \int_0^\infty e^{-x}(x - \alpha)\, dx = 1 - \alpha.$$

Hence $\alpha = 1$ and $\mathbf{e}_2 = (x - 1)$.

Then $\mathbf{e}_3 = x^2 - \beta - \gamma\mathbf{e}_2$ where β, γ are constants chosen so that $\mathbf{e}_3 \cdot \mathbf{e}_1 = \mathbf{e}_3 \cdot \mathbf{e}_2 = 0$. These conditions give

$$0 = \int_0^\infty e^{-x}x^2\, dx - \beta\int_0^\infty e^{-x}\, dx - 0$$

i.e. $0 = 2 - \beta$, and

$$0 = \int_0^\infty e^{-x}(x^3 - x^2)\, dx - 0 - \gamma\int_0^\infty e^{-x}(x - 1)^2\, dx$$

i.e. $0 = 4 - \gamma$.

Hence $\mathbf{e}_3 = x^2 - 2 - 4\mathbf{e}_2 = x^2 - 4x + 2$.

The polynomials we obtain in this way are the Laguerre polynomials. We have used the facts that

$$\int_0^\infty e^{-x}\, dx = 1$$

and for $n \geqslant 1$

$$\int_0^\infty e^{-x}x^n\, dx = n!$$

These can be easily proved using integration by parts to obtain the recurrence relation

$$\int_0^\infty e^{-x}x^n\, dx = n\int_0^\infty e^{-x}x^{n-1}\, dx.$$

Using

$$\|\mathbf{f}\|^2 = \int_0^\infty e^{-x}[f(x)]^2\, dx$$

and

$$\int_0^\infty e^{-x}x^n\, dx = n!$$

we obtain

$$\|\mathbf{e}_1\|^2 = \int_0^\infty e^{-x}\, dx = 1$$

$$\|\mathbf{e}_2\|^2 = \int_0^\infty e^{-x}(x - 1)^2\, dx = 1$$

$$\|\mathbf{e}_3\|^2 = \int_0^\infty e^{-x}(x^2 - 4x + 2)^2\, dx = 4$$

Hence the normalized vectors are $\{1, (x - 1), \frac{1}{2}(x^2 - 4x + 2)\}$.

LINEAR MATHEMATICS